Handwritten Notes For Botany...

NITROGEN FIXATION AND METABOLISM

I0481769

V.Darani M.Sc., M.Phil., SET

Chapters Enclosed....

Chapter - 1.

Nitrogen - its needs and modes of fixation

Nitrogen Metabolism refers to the synthesis and breakdown of nitrogen compounds in plants.

Nitrogen is one of the major elements required for plant growth. Nitrogen is an universally occurring element. Apart from water and mineral salts, the next major substance in plant cell is protein (about 10-12% of the cell). These Proteins are the building blocks of the protoplasm and are made up of nitrogenous substances called amino acids. The Amino acids are synthesized when inorganic nitrogen of the environment is converted to organic nitrogen inside the plant.

Nitrogen is also constituent element of many other important organic compounds like chlorophyll, cytochromes, alkaloids, vitamins and nucleic acid. Thus nitrogen plays a very important and fundamental role in Metabolism, growth, reproduction and heredity.

The productivity of plants is determined by the availability of nitrogen for their growth. Though gaseous atmosphere is composed of 80% dinitrogen (N_2), it cannot be used as such. This is because the triple bonded structure

N≡N, is extremely stable and chemically very unreactive because of the high bond energy of the molecule. Dinitrogen must be converted into highly oxidesed States (NO_3 or NO_2) or most reduced state (NH_3) for plant utilisation.

Nitrogen Metabolism involves a series of biochemical changes which result in the construction of complex nitrogenous substances from simple nitrogen containing substances and their breakdown into simple substances.

Anabolism is the term used to denote the synthesis of nitrogenous compounds and catabolism involves the breakdown of complex nitrogenous materials. The anabolic processes are otherwise known as constructive processes which includes nitrogen fixation and the synthesis of amino acids, Proteins and nucleic acids.

Catabolism is otherwise termed as destructive processes which includes denitrification, ammonification breakdown of nucleic acids and Proteolysis.

Both the anabolic and catabolic processes are going on side by side in plants continuously. The available or utilisable form of nitrogen are ammonia, nitrate, nitrite and organic nitrogen of which nitrates are the chief form of nitrogen readily taken by plants.

Role of nitrogen in plants

In 1999 Epstein stated that the plant tissues contain about 1.5 % nitrogen in their total dry weight

The important role played by nitrogen in plants
are as follows:

→ Plant growth

→ Nitrogen forms the constituent of amino acids,
amides, proteins, nucleic acids, nucleotides, coenzymes,
hexoamines, Chlorophylls etc.

→ Nitrogen containing vitamins serve as prosthetic
groups in some enzymes

→ The building blocks of protein are amino acids
whose structure is composed of nitrogen.

→ Proteins form the structural framework of cell and
enzymes that catalyze various biochemical reactions
Proteins form 10-12% of the total weight of the cell.

→ Since nitrogen is the constituent of many plant
cell components like amino acids, nucleic acids etc,
their deficiency will rapidly inhibit the growth
of plants

→ Since nitrogen is a constituent of nitrogenous bases
of nucleic acids, it takes part in determining the
heredity of plants

On the whole deficiency in the source of
nitrogen will lead to retarded growth and reproduction
in plants.

Sources of Nitrogen

⇒ Atmospheric nitrogen (Molecular nitrogen)

Although about 78% of the earth's atmosphere
— 2 —

is composed of nitrogen, the majority of plants can't utilize this form of nitrogen. Only some bacteria, blue green algae are capable of fixing atmospheric nitrogen in a form utilized by plants

⇒) Nitrates, Nitrites, Ammonia in the soil (Inorganic Nitrogen)

Among these nitrate is the chief form of nitrogen taken up by plants from soil

⇒ Amino acids (organic Nitrogen) in the soil

Many soil microorganisms utilize this form of nitrogen. Even some higher plants may also take this form of nitrogen.

⇒) Organic Nitrogenous compounds in Bodies of the Insects

Insectivorous plants fulfil their nitrogen requirements by catching small insects and digesting them.

| Nitrogen cycle |

The cycling of nitrogen between abiotic and biotic systems is called nitrogen cycle. It is a gaseous cycle.

Nitrogen is present both in plants and the outer environment and is continuously moving in and out of the living beings in nature.

The main reservoir of nitrogen is atmosphere. It contains 78% nitrogen. Plants require nitrogen for many vital activities and obtain nitrogen in the form of ammonium salts, nitrites and nitrates. These compounds are formed from atmospheric nitrogen by a process called nitrogen fixation.

The steps involved in nitrogen cycle are as follows:

a) Nitrogen fixation
b) Nitrogen Assimilation
c) Ammonification
d) Nitrification
e) Denitrification
f) Sedimentation

a) <u>Nitrogen fixation</u>

The conversion of free nitrogen of atmosphere into biologically acceptable form or nitrogenous compounds is referred as nitrogen fixation. This process is of two types:

→ Physicochemical nitrogen fixation (or) Non-biological nitrogen fixation

→ Biological Nitrogen fixation

→ <u>Physicochemical Nitrogen fixation</u>

Electrochemical nitrogen fixation is the fixation of nitrogen by the action of lightning.

→ <u>Biological nitrogen fixation</u>

It refers to the conversion of free N_2 into soluble salts by the activity of certain organisms. These organisms are called N_2 fixing organisms.

Eg:- <u>Rhizobium</u>, <u>Azotobacter</u>, <u>Clostridium</u>, <u>Bacillus</u> etc

Fig 1: Flow of Nitrogen.

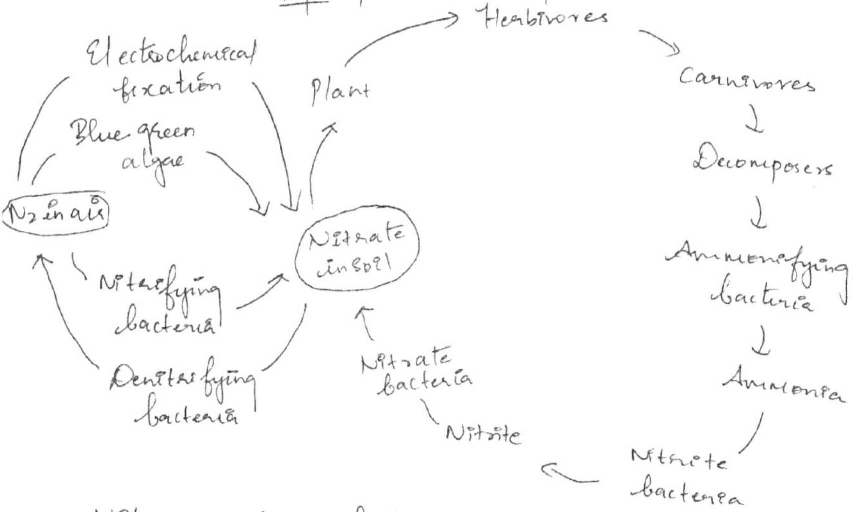

Nitrogen of the biotic system flows into the abiotic system by the following 6 methods:

☞ Proteolysis　　　　　☞ Ammonification

☞ Nitrification　　　　☞ Nitrate reduction

☞ Denitrification　　　☞ Combustion

☞ Proteolysis

It is the enzymatic hydrolysis of proteins in dead organic matter into amino acids. Microorganisms

The microorganisms produce two types of Proteolytic enzymes Proteinases and Peptidases. Proteinases convert protein to smaller units of Peptides. Peptidases convert Peptides into aminoacids. The end products of proteolysis are aminoacids. The overall reaction is summarised in figure 1.

$$\boxed{\text{Proteins}} \xrightarrow{\text{Proteinases}} \boxed{\text{Peptides}} \xrightarrow{\text{Peptidases}} \boxed{\text{Amino acids}}$$

b) **Nitrogen assimilation**

Inorganic nitrogen in the form of nitrates, nitrites and ammonia is absorbed by the green plants and converts into nitrogenous organic compounds. Nitrates are first converted into ammonia which combines with organic acids to form aminoacids. Amino acids are used in the synthesis of proteins, enzymes, chlorophylls, nucleic acids etc. Animals derive their nitrogen requirement from the plant proteins. Plant proteins are not directly utilised by animals. They are first broken down into amino acids during digestion and the amino acids are absorbed and manipulated into animal proteins, nucleic acids etc.

c) **Ammonification**

The dead organic remains of plants and animals and excreta of animals are acted upon by a number of microorganisms especially actinomycetes and bacilli. These organisms utilize organic compounds in their metabolism and release ammonia.

d) Nitrification

Fig 2: Nitrogen cycle in ecosystem

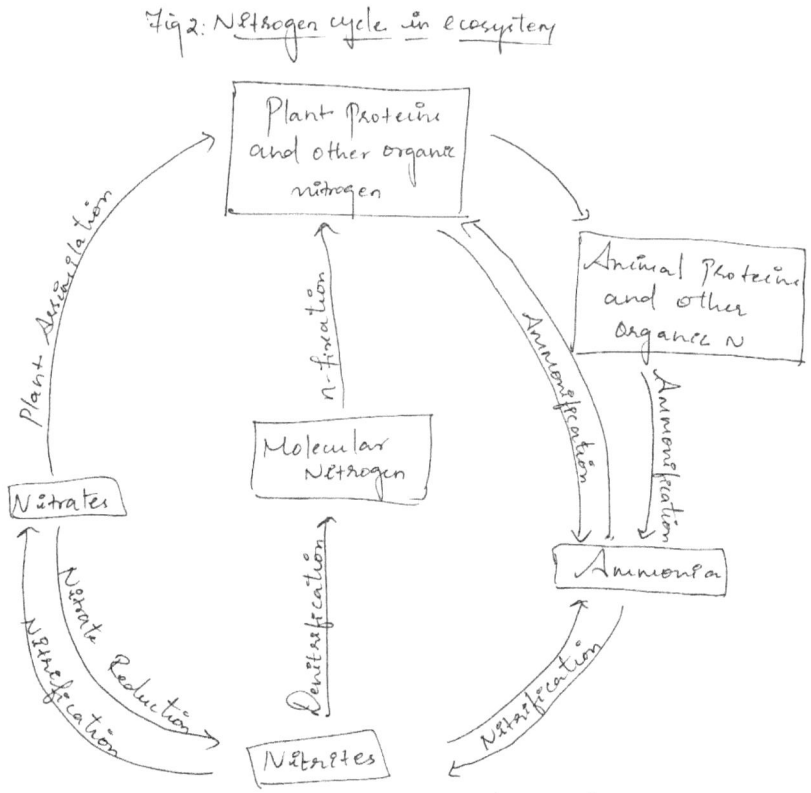

Certain microbes in ocean and soil (eg:- Nitrosomonas, Nitrococcus) convert ammonia into nitrites and then nitrites into nitrates. These microorganisms primarily use the energy of dead organic matter in their Metabolism.

$$2NH_4^+ + 2O_2 \longrightarrow NO_2^- + 2H_2O + Energy$$

Some nitrates are also made available through weathering of nitrate containing rocks.

e) Denitrification :

Conversion of ammonia and nitrates into free nitrogen by microorganisms is termed as denitrification. (eg:- Thiobacillus denitrificans, Micrococcus sp)

$$2NO_3^- \longrightarrow 2NO_2^- \longrightarrow 2NO \longrightarrow N_2O \longrightarrow N_2.$$

f) Sedimentation :

Nitrates of soil are washed down to the sea or leached deep into the earth along with percolating water. Nitrates thus lost from the soil surface are locked up in the rocks. This is the sedimentation of nitrogen. Nitrogen of rocks is released only when the rocks are exposed and weathered.

Thus a large part of nitrogen is fixed up and stored in plants, animals and microbes. Nitrogen leaves the living system in the same amount taken in from the atmosphere and the input and output of nitrogen is balanced in the ecosystem. The overall nitrogen cycle is given in the figure 2.

Non - biological nitrogen fixation

The nitrogen fixation in the nature without the involvement of living beings is called non-biological nitrogen fixation.

It includes 2 types
a) Atmospheric Nitrogen fixation
b) Industrial Nitrogen fixation

Atmospheric Nitrogen fixation

The conversion of molecular nitrogen into nitrates by natural forces is called Atmospheric nitrogen fixation. It takes place by lightning and photochemical reactions. Atmospheric nitrogen fixation accounts for about 10% nitrogen in the soil. Global nitrogen fixation is about 19×10^6 tonnes/year.

Lightning during thunder provides high energy for the physicochemical nitrogen fixation. About 8% of the total nitrogen fixed in the soil is contributed by lightening. The high energy discharged in the atmosphere produce highly reactive OH^- radicals, H^+ and O^- from water vapour (H_2O). The O_2 reacts with N_2 and form nitric oxide (NO). The NO reacts with O_2 to form nitrous oxide (NO_2)

$$N_2 + O_2 \xrightarrow{\text{high energy}} 2NO$$
$$2NO + O_2 \xrightarrow{\text{oxidation}} 2NO_2$$

During rain, the NO_2 combines with rain water to form nitrous/nitric acid.

$$2NO_2 + H_2O \longrightarrow HNO_2 + HNO_3$$

The nitrous/nitric acid comes to the ground along with rain water. In the soil nitric acid reacts with alkali radicals (Ca and K) to produce their nitrites and nitrates

$$\text{Ca or K salt} + HNO_3 \longrightarrow CaNO_3 \text{ or } KNO_3$$

Photochemical reactions account for the fixation of

about 2% of available nitrogen in the soil. Light energy from meteorite trails and cosmic radiation enables the atmospheric nitrogen to react with O_2 to form nitrous oxide (NO). This NO reacts with water molecules in the atmosphere to form nitric acid.

Industrial Nitrogen Fixation

The conversion of molecular nitrogen into a nitrogenous compound that can be readily used by plants by adopting a chemical method is called industrial nitrogen fixation. The usable compounds produced are ammonia, nitric acid, ammonium nitrate, urea, sulphur coated urea, isobutylidene diurea, ammonium sulfate, calcium nitrate, ammonium chloride or calcium cyanide. These compound are used as nitrogen fertilizers in the soil.

In Industries, ammonia is manufactured from nitrogen and hydrogen by using Haber-Bosch process. In this process, nitrogen obtained from the atmosphere is mixed with hydrogen obtained from electrolysis of water under high pressure with 200°C in a closed chamber. As a consequence, nitrogen combines with hydrogen to form ammonia (NH_2). Global industrial nitrogen fixation is estimated to be 80×10^6 tonnes/year

$$N_2 + 3H_2 \xrightarrow[\text{200 atmosphere}]{\text{200°C}} 2NH_3$$

Biological Nitrogen fixation

The conversion of molecular nitrogen into ammonia by microorganisms is known as biological nitrogen fixation. It takes place at usual atmospheric pressure and 20°C temperature by the aid of enzyme nitrogenase. Biological nitrogen fixation contributes 90% of fixed nitrogen of earth. The global biological nitrogen fixation is 175×10^6 tonnes/year.

$$N_2 + 3H_2 \xrightarrow{\text{Nitrogenase}} 2NH_3$$

Some 10% bacteria and blue green algae are capable of reducing the atmospheric nitrogen into ammonia in their cells. The process of nitrogen reduction is called diazotrophy or nitrogen fixation. The enzyme nitrogenase is useful in nitrogen fixation. Microbes performing the process of nitrogen fixation is called nitrogen fixers or diazotrophs. Ammonia produced during nitrogen fixation is readily available to Plants for direct use.

Green plants use ammonia to synthesize nitrogen containing compound such as arginine, asparagine, allantoin and allantoic acid. These nitrogen containing compounds synthesized directly from ammonia are known as ureides. Ureides are used in the Metabolism of nucleic acid and Proteins.

⟹ <u>Nitrogen Fixing Organisms</u>
* Free-living Diazotrophs.

Habit when diazotrophic	Genus or type	Examples of species
1) Strict anaerobes	Archaebacteria	Methanosaruna	M. bakeri
		Methanococcus	M. thermolithotrophicu
	Eubacteria	Bacillus	B. polymyxa
			B. macerans
		Clostridium	C. pasteurianum
			C. butyricum
		Desulfotomaculum	D. nurumi
			D. orientii
		Desulfovibrio	D. desulfuricans
			D. vulgaris
			D. gigas
2) Facultative anaerobes	Klebsiella	K. pneumoniae
			K. aerogenes
		Enterobacter	E. aerogenes
			E. cloacae
		Escherichia	E. coli
3) Micro-aerobes	Aquaspirillum	A. peregrenum
			A. fasciculus.
		Arthrobacter	A. fluorescens
		Azospirillum	A. lipoferum
			A. brasilense
4) Aerobes	Azotobacter	A. beijerinckii
			A. chroococcum
			A. vinelandii

	Beijerinckia	B. alexei
		B. indica
		B. mobiles
	Derxia	D. gummosa
5) Photosynthetic bacteria	Rhodopseudomonas	R. palustris
		R. capsulata
		R. viridis
	Rhodospirillum	R. rubrum
		R. tenue
		R. fulvum
	Chromatium	C. vinosum, C. minus
		C. weissei

6) Cyanobacteria
   (aerobic)
   Anabaena
   cylindrospermum
   Nostoc
   Calothrix

* <u>Symbiotic systems</u>

1) Rhizobium - Legume Associations
      (Legumes + Rhizobium)
2) Bradyrhizobium - Legume Associations
3) Bradyrhizobium - Non-legume Associations
      (Parasponia + B. parasponia)

4) Frankia (actinomycetes) - non legume associations
      non legumes (Alnus, Casuarina, Myrica) + Frankia

5) Azotobacter paspali, Azospirillum - tropical grasses
                                        Association

6) Cyanobacterial Associations

- With Angiosperms — Gunnera + Nostoc
- With gymnosperms — Cycad (Macrozamia communis) + Nostoc or Anabaena
- With Pteridophytes — Azolla (fern) + Anabaena azollae
- With Bryophytes — Anthoceros + Nostoc
- Lichens — Symbioses of fungi (ascomycete and basidiomycetes) with green algae and cyanobacteria

The genus Rhizobium is divided into two groups in Bergy's manual:

1) Fast growing Rhizobia — R. trifolei, R. leguminosarum, R. phaseoli, R. meliloti

2) Slow growing Rhizobia — R. japonicum, R. lupini

Non-symbiotic Nitrogen fixation

The nitrogen fixation process when carried out by microorganisms that are free-living or those that do not form symbiotic association is called non-symbiotic nitrogen fixation. It is of two types:

⇒ Nitrogen fixation by free living Autotrophs:

a) Aerobic eg:- Some cyanobacteria (blue-green algae).

All those blue green algae which can fix atmospheric nitrogen, usually contain heterocyst such as Nostoc, Anabaena, Tolypothrix, Aulosira, Calothrix etc. But all the heterocysts bearing blue green algae may not be atmospheric nitrogen fixers. A few non-heterocystous blue green algae such as Gloeotheca are also known to fix atmospheric nitrogen.

b) Anaerobic – eg:- Certain bacteria such as Chromatium and Rhodospirillum

⇒ Nitrogen fixation by free living heterotrophs

a) Aerobic – eg:- Certain bacteria such as Azotobacter, Azospirillum, Derxia, Beijerinckia

b) Facultative – eg:- Certain bacteria such as Bacillus and Klebsiella

c) Anaerobic – eg:- Certain bacteria such as Clostridium and Methanococcus.

The free living autotrophic diazotrophs synthesize their own food by photosynthesis using the sunlight or by chemicals. The free living heterotrophic diazotrophs on the other hand use dead organic matter as food and do nitrogen fixation.

Features favouring Non-Symbiotic nitrogen fixation

All diazotrophs contain enzyme nitrogenase which catalyzes the reduction of N_2 into NH_3. Nitrogenase is sensitive to oxygen, hence it prefers anaerobic conditions for nitrogen fixation. But in microbes, the oxygen level is usually high. The high oxygen level leads to oxidation of nitrogenase and hence that enzyme become inactive.

Non-Symbiotic nitrogen fixers have the following special features for nitrogen fixation:

→ **Spatial** separation of Nitrogen fixing cells:

Blue green algae are photoautotrophic diazotrophic living on moist soil. They have Photosystem I and II, so that Oxygen is evolved during light reaction of Photosynthesis. This oxygen may inhibit the nitrogenase activity. In these algae, Photosynthesis is restricted to Vegetative cells and nitrogen fixation is restricted to heterocysts. Eg:- Anabaena, Nostoc, Tolypothrix

Fig:- Anabaena Trichome

Heterocyst   vegetative cell

Fig: Diagram denoting Spatial separation of Nitrogen fixation

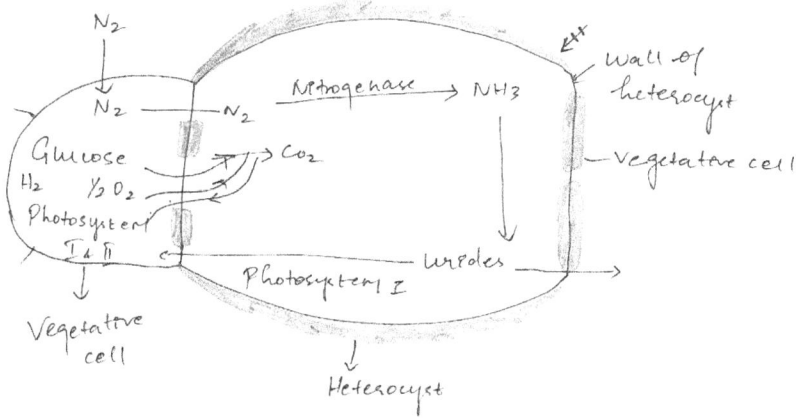

Heterocysts are thick walled cells which lack Photosystem II due to which oxygen will not be evolved during photosynthesis. The thick wall of the heterocysts prevent the entry of oxygen from adjacent cells into the heterocysts. But inside the heterocysts, respiration will go on continuously. Because of these conditions, Partial anaerobic condition develop in heterocysts which enhance nitrogen fixation. Heterocysts have various enzymes needed for anabolism of urides.

In non-heterocystous cyanobacteria, nitrogen fixation occurs in some internal cells in which respiratory oxygen consumption is relatively high; this higher rate of respiration brings out low oxygen level for nitrogenase activity.

→ <u>Protein Nitrogenase Association</u>

Azotobacter, Derxia and Mycobacterium are aerobic heterotrophic nitrogen fixing bacteria. In these bacteria, some intracellular proteins are found associated with nitrogenase, when oxygen level is high in the cells. When the oxygen level comes down, the protein release the nitrogenase free; the free nitrogenase fixes nitrogen as usual.

→ <u>High Rate of Respiration</u>

Some aerobic heterotrophic diazotrophs show high rate of respiration, so the cellular oxygen level is reduced to certain extent. At the reduced oxygen level, nitrogenase reduces N_2 into Ammonia

→ Time Specific Nitrogenase Activity

Some photoautotrophic aerobic diazotroph like Rhodopseudomonas capsulata does photosynthesis in the day time and nitrogen fixation in the night. In the night, oxygen level comes down in the cell due to continuous respiration.

→ Association with rapid oxygen consumers :

Some microorganisms cannot fix nitrogen when they are living away from the rhizosphere zone of some higher plants. Eg:- Klebsiella, Rhodopseudomonas etc, On the other hand when they dwell in the rhizosphere of high oxygen consuming plants, they fix atmospheric nitrogen. Here, plant roots create a partial anaerobic condition that favours nitrogen fixation.

→ Presence of Hydrogenase

When nitrogen is not present, nitrogenase reduces hydrogen ions (H+) into H_2. It utilises reductants and ATP for this purpose which is a useless process. The hydrogen so produced, while reaching higher concentration, inhibits nitrogen reduction

$$2H^+ + 2e^- + 4ATP \xrightarrow[Mg^{2+}]{Nitrogenase} H_2 + 4ADP + 4iP$$

Some diazotrophs (eg:- Azobacter chroococcum and Anabaena cylindrica) contain the enzyme hydrogenase. Hydrogenase transfers electrons to high energy potential electron acceptor and combines ½ O_2 and 2H+

into H_2O. This process is termed hydrogen uptake. The hydrogenase performing this function is known as uptake hydrogenase or irreversible hydrogenase.

Hydrogen uptake reduces the H_2^- and O_2^- levels in the nitrogen fixing cells. Further, it supplies ATPs to provide energy for nitrogen reduction and reduces the wastage of reductant. Therefore, nitrogenase is safe in the nitrogen fixing cell and does nitrogen reduction.

$$H_2 + \frac{1}{2}O_2 \xrightarrow{\text{Hydrogenase}} H_2O + Energy$$

→ colonization

The unicellular blue green alga Gleocapsa forms a globular colony by aggregation of many cells. Outer cells of the colony prevent the entry of oxygen into the inner cells. Hence, the inner cells are slightly anaerobic to do nitrogen fixation.

Nitrogenase

Biological nitrogen fixation is catalyzed by the enzyme nitrogenase. Nitrogenase consists of one larger subunit and one smaller sub unit. The larger sub unit is called molybdenum ferrous protein (Mo-Fe protein) or nitrogenase reductase. The smaller subunit is called ferrous protein.

Molecular weight of Mo-Fe protein is 200,000 - 245,000 daltons. It is made up of two identical large α-polypeptide chains and two identical small β- polypeptide chains. The α- and β- chains are bound together by 2 Mo atoms, 12-32 Fe atoms and 30 sulphur atoms. An Mo atom is surrounded by 7 ferrous atoms, 9 sulphur atoms and 1 homocitrate. Thus α-peptide contains 2Fe : 4S cluster and β-peptide has Mo : 3Fe : 3S cluster. Thus there are two metal centres in these subunits. Nitrogenase reductase is a cold tolerant protein. It is inactive at 0°C

The molecular weight of ferrous protein is 60,000~60,700 daltons. It is also referred as dinitrogenase reductase. It consists of two more or less identical polypeptide chains which are held together by 4 ferrous atoms and 4 sulphur atoms, forming one 4Fe : 4s cluster. It carries Mg- ATP for nitrogen reduction. These two subunits are held together by two S-S linkages between the peptides. The Fe-Protein acts as a redox site for nitrogen reduction.

Fig: Nitrogenase complex - components

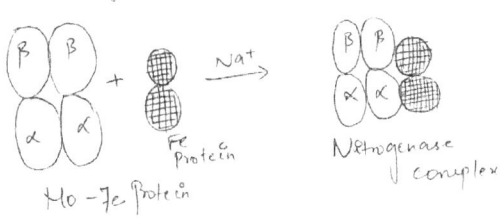

Mo - Fe Protein   Nitrogenase complex

In the presence of Na+ ions, Mo-Fe protein and Fe-Protein combine together to form an active nitrogenase complex. They readily dissociate after the reduction of nitrogen. They occur in the ratio of 1:2 in nitrogen fixing cells.

Mg-ATP serves as the nitrogenase reduction system to supply energy and to donate electrons needed for the reduction of nitrogen.

Chapter - 2

Mechanism of Nitrogen Reduction

Requirements for Nitrogen Fixation

Energy source and electron donors are required by nitrogenase for the purpose of reducing nitrogen.

⇒ **Energy source :—**

ATPs released during Carbohydrate, Protein and lipid Metabolisms react with Mg^{++} ions to form Mg - ATPs. The Mg^{++} of Mg-ATP binds with a Fe - Protein to form an active complex. This Mg-ATP is hydrolyzed into Mg-ADP and inorganic phosphate (iP) to supply energy. About 12-15 Mg - ATPs are required to reduce one molecule of N_2 into NH_3

$$Mg - ATP + H_2O \longrightarrow Mg - ADP - iP;$$
$$\Delta G^\circ = -7.3 \ K.cal/mol$$

$$N_2 + 6H^+ + 6e^- \xrightarrow[12 Mg-ADP+ 12 \ iP]{12 \ Mg-ATP} 2NH_3$$

⇒ **Electron source :—**

The electrons needed for the reduction of nitrogen are provided by electron donors or reductants. Ferredoxin serves as the electron donor in many diazotrophs. In Azotobacter and Blue green algae, NADPH functions as the electron donor. In anaerobic diazotrophs, Pyruvate transfers electrons to nitrogenase complex for reducing nitrogen

Mechanism of Beological Nitrogen Fixation

During biological nitrogen fixation, gaseous nitrogen (N_2) is reduced into ammonia (NH_3) by the enzyme nitrogenase. The steps involved in the process are summarised as follows:

$$\boxed{\text{Fe-protein} \xrightarrow{\text{6e}^-} \text{Reduced Fe-protein}}$$

The Fe Protein (dinitrogenase reductase) is reduced after receiving electrons from ferredoxin (or NADPH)

$$\boxed{\text{Reduced Fe-protein} + 12 Mg\text{-}ATPs \longrightarrow \text{Reduced Fe protein Mg-ATPs complex}}$$

The reduced Fe-Protein accepts 12 molecules of Mg-ATP and forms a reduced Fe-Protein-Mg-ATPs comple (RFP-MA Complex). The Mg^{++} ions activate the Fe-protein

$$\boxed{\text{Nitrogenase} + N_2 \longrightarrow \text{Nitrogenase Nitrogen complex}}$$

The nitrogenase (MO-Fe protein) accepts a molecule of nitrogen and gets converted into Nitrogenase nitrogen complex.

RFP-MA Complex + NNC $\longrightarrow$ Nitrogenase complex

RFP-MA complex combines with nitrogenase nitrogen complex to form an active nitrogenase complex in the

Presence of Na^+ ions. Electrons in the RFP-MA complex is transferred to nitrogenase for reducing nitrogen. During this electron transfer, reduction of $2H^+$ ions to H_2 May takes place.

The reduced nitrogenase in the complex accepts $6H^+$ ions from cytoplasm and reduces nitrogen into ammonia using 6 electrons. The electrons present in Fe atoms of nitrogenase are used for this purpose. The reduction of nitrogen involves the following three steps:

$$NEN + 2H^+ \xrightarrow{2e^-} HN=NH$$

N_2 reacts with $2H^+$ ions by consuming 2 electrons to form a diamide

$$HN=NH + 2H^+ \xrightarrow{2e^-} H_2N-NH_2$$

Diamide reacts with $2H^+$ ions by consuming 2 electrons to form a hydrozine

$$H_2N-NH_2 + 2H^+ \xrightarrow{2e^-} 2NH_3$$

Hydrozine reacts with $2H^+$ ions by consuming 2 electrons to form 2 molecules of ammonia.

After the reduction of nitrogen into ammonia, the nitrogenase complex dissociates into a Fe-Protein, nitrogenase, Mg^{++} and ADPs. NH_3 so Produced is released in the cytoplasm. The enzyme is now available to reduce another molecule of nitrogen.

Electron transfer during Nitrogen fixation — A glance

Direct Measurement of nitrogen fixation can be done by using Mass spectroscopy.

The electrons are transferred from reduced ferredoxin or flavodoxin or other effective reducing agents to Fe-Protein component which gets reduced. From reduced Fe-protein, the electrons are given to Mo-Fe protein component which in turn gets reduced and is accompanied by hydrolysis of ATP into ADP and inorganic phosphate (Pi). Two Mg^{++} and 2 ATP molecules are needed per electron transferred during this process.

From the reduced MO-Fe protein, electrons are finally transferred to molecular nitrogen (N_2) and 8 Protons, so that two ammonia and one hydrogen molecule are produced (fig)

Fig: Mechanism of biological nitrogen fixation

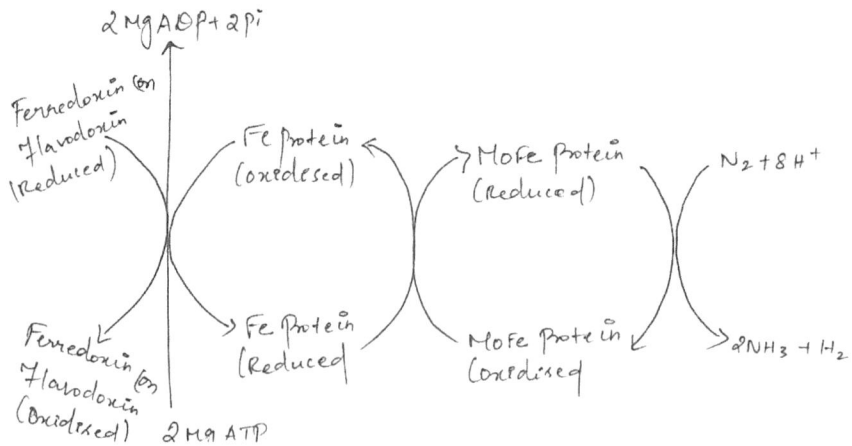

It might be expected that 6 electrons and 6 protons are needed for reducing one molecule of N_2 into 2 molecules of ammonia (NH_3). But, the reduction of N_2 is obligatorily linked to the reduction of 2 protons to form one H_2 molecule. It is believed that this is needed for the binding of nitrogen at the active site.

The electrons for regeneration of reduced electron donors (ferredoxin, flavodoxin etc) are provided by the cell Metabolism eg:- Pyruvate oxidation

Substantial amount of energy is lost by the microorganisms in the formation of H_2 molecule during nitrogen fixation. However, in some rhizobia, hydrogenase enzyme is found which splits H_2 to electrons and protons ($H_2 \rightarrow 2H^+ + 2e^-$). These electrons May then be used again in reduction of nitrogen, thereby increasing the efficiency of nitrogen fixation.

Chapter - 3

Symbiotic Nitrogen Fixation

Fixation of nitrogen by microorganisms that are associated with the roots of plants is termed as symbiotic nitrogen fixation.

The microorganisms exhibiting symbiotic association with plants and perform nitrogen fixation are termed as symbiotic nitrogen fixers or symbiotic diazotrophs. The diazotrophs derive nourishment from plants and in return supply nitrogen to plants.

During symbiotic association, some microorganisms induce nodule development in plant roots. This type of symbiosis is called rhizocoenosis. Eg:- Rhizobium, Frankia The nitrogen fixation in this case occur in the root nodules.

Host specificity

Although many species of Rhizobia live in the rhizosphere of a legume, a particular species alone can establish symbiotic association with its roots. This selective infection of Rhizobium on specific plants is called host specificity. For example,

Rhizobium leguminosarum establishes root nodules in Pea.

Rhizobium phaseoli induces nodulation on roots of beans

Rhizobium japonicum forms root nodules in soybeans

The host specific infection of Rhizobium depends upon the specific flavonoid secreted by the roots of legumes. Alfalfa exudes luteolin, a white clover exudes dihydroxy-flavone along with lectins. The root exudate induces certain genes of a Particular species of Rhizobium to produce a host determinant compound on its cell wall.

In most cases, the host determinant compound is a capsular polysaccharide (cp). The lectin produced by the legume root has affinity to the capsular polysaccharide. Therefore, it binds with the capsular polysaccharide of the specific Rhizobium species and the other end of the lectin binds with cell wall polysaccharide of a root hair. Here, lectin acts as a bridge molecule. After recognition, the Rhizobium infects the root hair.

Fig:- Mechanism of host specific infection

Formation of Root nodules in Legumes

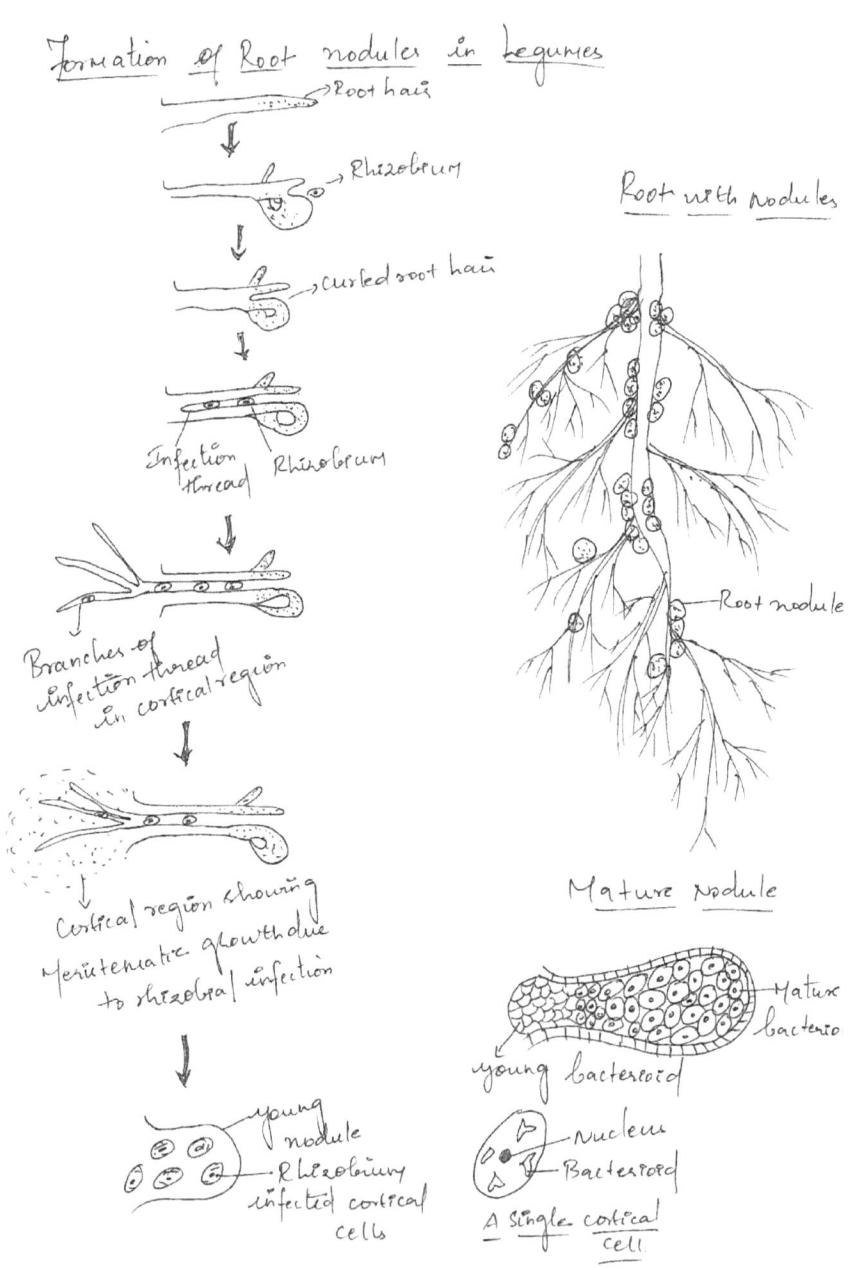

→ Root hair

→ Rhizobium

→ curled root hair

Infection thread — Rhizobium

Branches of infection thread in cortical region

Cortical region showing Meristematic growth due to rhizobial infection

young nodule — Rhizobium infected cortical cells

Root with nodules

— Root nodule

Mature Nodule

— Mature bacterio

young bacteroid

— Nucleus
— Bacteroid

A single cortical cell

Rhizobia occur as the free living organisms in the soil before infecting the respective host plants to form root nodules. The symbiosis between rhizobia and leguminous host plant is not always obligatory. However, under conditions of limited nitrogen supply in the soil, there is elaborate exchange of signals between the two symbionts for development of symbiotic relationship.

There are separate host specific genes and rhizobial specific genes which are involved in nodule formation. The host plant genes are called as nodulin or nod genes while rhizobeal genes are called as nodulation or nod genes. Some Nod factors produced by rhizobia act as signals for symbiosis.

The rhizobia migrate and accumulate in the soil near the roots of the legume plant in response to the secretion of certain chemicals like flavonoids and betaines by the roots. Root hairs of legumes produce specific sugar binding proteins called lectins. These lectins are activated by Nod factors to facilitate the attachment of rhizobia to the root hairs lips inturn become curved.

Rhizobia now secrete enzymes which degrade the cell walls of root hairs at the point of their attachment for entry into the root hair. From root hairs, the rhizobia enter into the cells of inner layers of cortex through infection threads (tubular extensions of the unfolded plasma membrane produced by fusion of Golgi-derived membrane Vesicle). The rhizobia continue to multiply inside infection thread and are released into cortical cells in large numbers, where they cause Cortical cells to multiply and ultimately result in the formation of nodules on the upper surface of the root. After their release into cortical cells, the rhizobia stop dividing and enlarge.

Electron microscopic studies have shown groups of rhizobia to the surrounded by single Membranes which originate from host cell plasmamembrane. The enlarged and non motile groups of bacteria inside the Membranes are called as bacteroids and the membrane surrounding them as Peribacterial Membrane. The space between bacteroids and peribacteroid Membrane is called as Peribacteroid space. These bacteroids are aerobic and the nitrogenase enzyme is found in them.

The bacteroids lack a firm wall and are osmotically labile. In root nodule cells of Glycine max often groups of 4-6 bacteroids are enclosed inside the peribacteroid Membrane.

The number of chromosome in cortical cells infected by rhizobia which later develop into nodule is double the number of chromosome in other somatic cells of the legume (ie. they are tetraploid) and seems to be pre-requisite for nodule formation.

Apart from infected cells which are tetraploid, some uninfected deploid cells are also found in nodule. The nodule has its own vascular system which is connected with vascular system of the root to facilitate transfer of fixed nitrogen i.e. NH_3 to the host and carbohydrates and other nutrients from host to the bacteroids

In root nodules of leguminous plants, a red pigment an oxygen binding heme protein which is very much similar to hemioglobin of red blood corpuscles is found. This pigment is called as leghaemoglobin and occurs in cytosol of infected nodule cells. Leghemioglobin gives pinkish-red colour to the nodules. The globin part of this pigment is synthesized in host plant genome in response to the bacterial infection, while its heme portion is synthesized by bacterial genome.

Although a correlation has been found between the concentration of hemioglobin and the rate of nitrogen fixation, but this pigment does not play a direct role in nitrogen fixation. It protects the nitrogenase inside the bacteroids from deterrement effect of oxygen and maintains adequate supply of oxygen to the bacteroids, so that through respiration ATPs continue to be generated which are required for nitrogen fixation.

After its formation inside bacteroids, ammonia (or NH_4^+) is released into cytosol of infected nodule cells where it is converted into amides (chiefly asparagine and glutamine) or ureids (chiefly allantoic acid, allantoin and citrulline). These amides or ureids are then translocated to shoots of host plant through xylem, where they are rapidly catabolized to NH_4^+ for entry into mainstream of ammonium assimilation

Factors Controlling nodulation

Environmental factors like high dose of nitrogen fertilizers, low concentration of CO_2 and dense population of other species of bacteria in the rhizosphere decrease the nodulation. The temperature between 25-30°C favours nodule formation. High intensity of light and high concentration of CO_2 inhibit nodulation.

Mechanism of nitrogen Fixation

As nitrogenase is sensitive to oxygen, root nodules of legumes have two mechanisms to protect nitrogenase from oxygen. They are as follows:

a) Oxygen Transport by Leghaemoglobin

Leghaemoglobin is a red, myoglobin like Protein present only in healthy root nodules of legumes. It is present outside the bacteroid, but in close contact with it

Since leghaemoglobin has a haem prosthetic group, it acts like haemoglobin in blood. It acts only as an oxygen carrier within the nodule and has no affinity with oxygen.

Leghaemoglobin combines with oxygen to form oxyleghaemoglobin and provides oxygen to plant cells for respiration. This reduces the level of oxygen around the bacterioid. It also supplies low level of oxygen to bacterioid that does not affect the activity of nitrogenase. Thus leghaemoglobin acts as a buffering system.

b) <u>Utilization of oxygen by hydrogenase</u>

Some Rhizobial strains possess hydrogenase, an enzyme that combines H_2 and O_2 to form H_2O (water). It also removes oxygen from the vicinity of nitrogenase in the bacteria. Meantime, it regenerates some ATPs lost by hydrogen reduction during nitrogen reduction. Thus hydrogenase makes a suitable microenvironment for nitrogenase activity.

<u>Requirements for Symbiotic Nitrogen reduction</u>:

Enzyme — Nitrogenase

Bacteroids synthesize ATPs, Proton (H^+), electron donors such as $NADPH_2$ and ferredoxin by oxidizing the sugars. From reduced ferredoxin electrons flow to MO-Fe protein

The enzyme nitrogenase complex receives energy from Mg-ATPs by hydrolysis. Mo-Fe protein reduces N_2 to NH_3 by using the electrons. Atleast 6 electrons are used to reduce one molecule of N_2 into two molecules of ammonia (NH_3). The Mo-Fe protein and Fe Protein Then separate from each other.

Steps involved in N_2 reduction are as follows:

→ | Fe-protein $\xrightarrow{6e^-}$ Reduced Fe-Protein |

Fe-Protein (dinitrogenase reductase) receives electrons from ferridoxin (or NADPH) and gets reduced.

→ | Reduced Fe protein + 12 Mg-ATPs → Reduced Fe Protein - Mg-ATPs complex |

Reduced Fe-Protein accepts 12 molecules of Mg-ATP and become a reduced Fe-Protein-Mg-ATP complex (RFP-MA complex). The Mg^{++} ions activate the Fe-Protein.

→ | Nitrogenase + N_2 → Nitrogenase nitrogen complex |

Nitrogenase (Mo-Fe Protein) accepts a molecule of N_2 and gets converted into a Nitrogenase nitrogen complex (NNC).

→ RFP-MA complex + NNC → Nitrogenase complex

RFp- MA Complex binds with nitrogenase nitrogen complex and form an active nitrogenase complex in the Presence of Na^+ ions. Electrons in the RFp- MA Complex is transferred to nitrogenase for reducing nitrogen. During this electron transfer, $2H^+$ ions May be reduced to H_2

→ Reduced nitrogenase accepts $6H^+$ ions from the cytoplasm and reduces N_2 into ammonia using 6 electron. The electrons Present in Fe atoms of nitrogenase are utilized for this Purpose. Reduction of nitrogen takes Place in 3 steps:

$$N \equiv N + 2H^+ \xrightarrow{2e^-} HN = NH$$

Nitrogen reacts with $2H^+$ ions by consuming 2 electro to form dianide

$$HN = NH + 2H^+ \xrightarrow{2e^-} H_2N - NH_2$$

Dianide reacts with $2H^+$ ions by consuming 2 electron to form hydrozine

$$H_2N - NH_2 + 2H^+ \xrightarrow{2e^-} 2NH_3$$

Hydrozine reacts with $2H^+$ ions by consuming 2 electrons to form 2 molecules of ammonia.

After the reduction of N_2 into NH_3, the nitrogenase complex dissociates into Fe-Protein, nitrogenase, Mg^{++} and ADPs. NH_3 so produced is released into the cytoplasm.

The enzyme is now available to reduce another molecule of nitrogen.

$$N_2 + 16 ATP + 8e^- + 10 H^+ \xrightarrow{Mg^{++}} 2 NH_4 + H_2 + 16 ADp + 16 ip$$

Fig: Overall process of N_2 fixation in a bacteroid of Rhizobium

Factors controlling Biological Nitrogen fixation

The factors that control biological nitrogen fixation are as follows:

➡ Environmental factor:

The factors that favour photosynthesis like adequate moisture, warm temperature, bright sunlight and higher CO_2 levels, also enhance biological nitrogen fixation. Rate of biological nitrogen fixation is usually maximum in the afternoon when sucrose (Carbohydrates) are rapidl

translocated from shoots to root nodules through phle
At this time, transpiration is also rapid which enhance
the translocation of fixed nitrogen from root nodules
and roots to shoots through xylem.

☞ Stage of Plant growth

In leguminous plants, maximum nitrogen fixation
occurs after flowering when demand for nitrogen increases
for developing fruits and seeds. These legumes have high
protein content in their seeds. In such legumes, about
90% of nitrogen fixation occurs during reproductive
stage and 10% of it occur in first two months of
vegetative stage

Rate of nitrogen fixation is higher in perennial
legumes (Alfalfa) in comparison to annual legumes
(Soybean) This is because in annual legumes, root nodules a
to be formed a fresh every year, whereas in perennial
legumes, the nodules are formed from the beginning of
the plant growth which persist in subsequent years, hence
can fix more nitrogen

☞ Fertilizers in Soil

Increased amount of NO_3^- and NH_4^+ in the
soil suppress formation of root nodules and cause
more rapid senescence of already established nodules.

☞ Genetic factors:

Many genetic factors control nitrogen fixation and

yields of leguminous plants :-

The host specific nodulin genes and rhizobial nodulation genes are responsible for nodule formation. Attempts are made to increase the efficiency of nodules formation by altering the rhizobial genes and selecting compatible varieties of host plants.

Reduction of N_2 to two molecules of NH_3 by nitrogen fixing organisms is coupled with the production of H_2 molecule causes wastage of energy. The enzyme hydrogenase found in some rhizobia and free living bacteria can split H_2 to electrons and protons,

$$H_2 \rightleftharpoons 2H^+ + 2e^-$$

Through genetic engineering techniques, bacterial strains can be developed with active hydrogenase enzy and consequently the rate of nitrogen fixation can be enhanced

Studies of free living bacterium <u>Klebsiella</u> pneumoni and certain free living species of <u>Rhizobium</u> have reveal the process of nitrogen fixation is controlled by a set of genes which are named as Nif genes. In these organisms, such genes are repressed while in rhizo associated with root nodules, they are usually derepressed even in the presence of NH_4^+.

Chapter - 4

Ammonia Assimilation and Reduction of Nitrate into Ammonia in Plants

≡ Assimilation of Ammonia

Ammonia produced in the bacteriroids diffuses into plant cells of the root nodule. Ammonia is used in the biosynthesis of urides such as glutamine, glutamate and aspartate

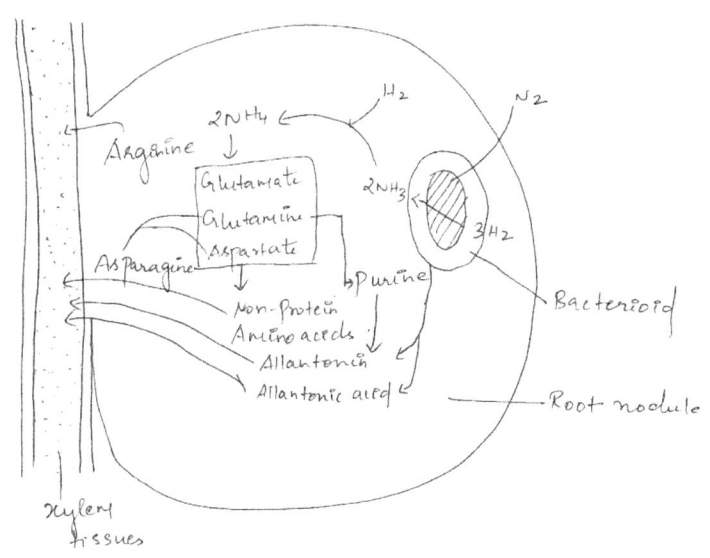

Fig. Diagram explaining the nitrogen fixation and uride Metabolism in the root nodule of legume

Purines are synthesized from glutamine and are used in the biosynthesis of allantoin and

allantoric acid. These compounds are released in the xylem sap of the root. Glutamine and aspartate are converted into asparagine and is released into the root

Some non-protein amino acids like homoserine, citrulline, canavanine etc, are synthesized from glutamate, glutamine and aspartate and released into the xylem sap.

All these compounds are transported to various parts of the plant where they are used in cellular Metabolism.

3 Reduction of Nitrate into Ammonia in Plants

Nitrates are the chief nitrogen compounds available in the soil naturally by means of non-biological nitrogen fixation. Plants absorb nitrates from the soil through the cells of the root.

Inside the cell, the nitrates are converted into ammonia. Since nitrates that are in highly oxidesed state are reduced to ammonia, this process is called Reduction of nitrate. It is also known as nitrate assimilation. It takes place in the cytoplasm of plant cells.

Nitrate reduction occurs in a series of reactions catalyzed by specific enzymes. The nitrate (NO_3^-) has five positive charges on it. It receives two electrons and release an oxygen atom to become a nitrate (NO_2). Then the nitrite is reduced into ammonia by accepting two electrons at three subsequent steps.

Thus, the resulting ammonia has three negative charges. The overall reaction of nitrate reduction is given below:

$$NO_3 \rightarrow NO_2 \rightarrow H_2 H_2 O_2 \rightarrow NH_2OH \rightarrow NH_3$$

Nitrate   Nitrite   Hyponitrite   Hydroxyl amine   Ammonia.

* Reduction of Nitrate into Nitrite

Plants assimilate nitrate in the root and leaf cells that contain the enzyme nitrate reductase. If the nitrate is available in small amounts it is assimilated in the root cells, if it is present in higher amounts, it is carried to the leaves through xylem and it is assimilate there.

The nitrate reductase enzyme consists of two identical subunits, each of which is a single polypeptide chain with three domains. FAD binding domain, heme binding domain and molybdenum complex binding domains are the 3 domains. The prosthetic groups such as FAD (Flavin adenine dinucleotide), heme and molybdenum are found attached to the respective binding domains.

There are two forms of nitrate reductase in plants. The most common form of nitrate reductase needs NADH as the electron donor and it is found in green cells. The other form uses NADPH as the reducing power and it is seen in non-green tissues in roots and shoots.

NADPH or NADH binds with FAD-binding domain of nitrate reductase and provides two electrons for the reduction of nitrate

$$NO_3^- + NAD(P)H^+ + H^+ + 2e^- \xrightarrow[\text{reductase}]{\text{Nitrate}} NO_2^- + NAD(P)^+ + H_2O$$
Nitrate   Nitrate

        Steps involved in the reduction of nitrate to nitrite are as follows:

→ cytochrome b_{557} reductase binds with nitrate reductase by a molybdenum binding sequence.

→ The NADPH or NADH binds with FAD-binding domain of nitrate reductase and supplies two electrons needed for the reduction of nitrate.

→ Electrons are transferred from NADPH or NADH to oxidized ferredoxin (FAD) and hence the FAD becomes reduced ferredoxin (FADH₂)

→ From FADH₂ electrons move to oxidized cyt b_{557}. At that time, two protons (H⁺) are released free and cyt b_{557} is reduced.

→ The reduced cyt b_{557} offers its two electrons to oxidized molybdenum atom in the nitrate reductase. As a result, the Mo gets reduced.

→ The reduced Mo sends its two electrons to nitrate (NO₃) and hence a bond between N and O atoms is broken down and one oxygen atom is released free. As a result, the nitrate is reduced to nitrite (NO₂)

→ Oxygen atom combines with the two protons to form a molecule of water.

Fig: Reduction of nitrate into nitrate in the cytoplasm

A notable exception to this mode of nitrate reduction is that in blue green algae. reduced ferredoxin acts as an electron donor instead of NADPH or NADPH. Therefore, water molecule has not been released during nitrate reduction

Nitrate reductase in higher plants is an inducible system which is synthesised de nova in the cells when the nitrate is present in the cell and is destroyed when the nitrate level comes down in the cell.

Nitrate Reductase activity is greatly decreased when plants are transferred from light to dark conditions or transferred to low CO_2 level. In the darkness, some Protein kinases phosphorylate the serine residues of nitrate reductase in the presence of Mg^{++} ions, so that 14-3-3 inhibitor (nitrate reductase inhibitory protein) can bind with serine. Hence, the nitrate reductase become inactivate

* Reduction of Nitrite to Ammonium

The conversion of nitrite into ammonium (NH_4) is called nitrite reduction.

Nitrite is highly reactive and toxic to cells. Therefore, it is readily transported from cytoplasm to chloroplasts (in leaf) or plastids (in root), where it is reduced to ammonium. The reduction of nitrite into ammonium is catalysed by the enzyme nitrite reductase

$$NO_2^- + 6Fd_{red} + 8H^+ + 6e \xrightarrow[\text{Reductase}]{\text{Nitrite}} NH_4^+ + 6Fd_{ox} + 2H_2O$$

Nitrite reductase of chloroplasts is slightly different from that of plastids in root cells. However, the general structure is the same in both cases. It consists of a single polypeptide chain containing an iron-sulfur cluster ($Fe_4 S_4$) and a specialized heme as the prosthetic groups. These prosthetic groups together form an active site for the binding of nitrite. Electrons from ferredoxin flow through iron-sulphur cluster and hence to reduce nitrite into ammonium

Fig :- Nitrate reduction in green cells

In leaf cells, ferredoxin is reduced by electrons coming from light reaction of photosynthesis

Chapter -5
Biochemistry of Nitrogen Fixation

The reduction reaction Catalysed by the enzyme can be represented as follows:

$$N_2 + 8H^+ + 8e^- \xrightarrow[16MgATP]{16MgADP + 16Pi} 2NH_3 + H_2 \uparrow$$

The evolution of H_2 concomitant to reduction of nitrogen is an inherent property of enzyme nitrogenase under normal conditions, for two moles of ammonia formed one mole of H_2 is evolved.

The basic requirements for biological fixation of dinitrogen are:

→ Nitrogenase enzyme complex

→ Mg^{2+} and ATP

→ a strong reducing agent and

→ anaerobic conditions

A.) Nitrogenase

Nitrogenase enzyme is also known as reduced ferredoxin (flavodoxin) : dinitrogen oxedoreductase (ATP-hydrolysing). The enzyme is composed of 2 oxygen sensitive non-haem iron proteins. The nitrogenase proteins can be separated into 2 brown protein fraction component 1 and component 2 which are necessary for nitrogen fixation

The larger protein, component-1, contains Mo, Fe and acid-labile sulphur and is called molybdenum iron protein (Mo-Fe Protein) or molybdoferredoxin or azofermic. Smaller protein, component 2, contains Fe and acid labile sulphur and is called Iron Protein (Fe-protein) or azoferredoxin or azofer.

Fe-Protein serves as one-electron donor and a specific reductase for Mo-Fe Protein. Mo-Fe Protein binds and reduces N_2 or other substrates. Hence, Fe-Protein has been designated as 'dinitrogenase reductase' and Mo-Fe Protein has been designated as 'dinitrogen reductase' or 'true dinitrogenase'.

⇒ **Fe-Protein or Component 2**

Fe Protein is a homomeric dimer and consists of 2 identical subunits (α_2). All Fe-Proteins contain 4 Fe and $4S^{2-}$ per mol. The Fe-Protein has 2 binding sites 1 for Mg ATP and another for Mg ADP. Its molecular weight varies from 55.5 kd for Bp2 to 72.6 kd for Xa2. This protein is extremely sensitive to oxygen and has a half life ($t_{1/2}$) in air of 30-45 s.

⇒ **Mo-Fe Protein or Component 1**

Mo-Fe protein is heteromeric tetramer of $\alpha_2 \beta_2$ subunit type. Thus it has 4 subunits of 2 different types. It also contains 2 Iron-molybdenum cofactors (Fe.Mo-co) per mol. This cofactor is probably the site of nitrogen reduction. The molecular weights ranges from 200 kd to 270 kd. The Mo-Fe Protein is also sensitive to oxygen and its half life in air is 10-24 Min.

⇒ **Third component Protein in Nitrogenase**

In some cases (eg - Rhodopseudomonas sp, Azospirillum) there is a genuine requirement for a protein that activates Fe-Protein. The inactive Fe-Protein require the enzyme together with ATP and Mn^{2+} to activate it. A protein component has been discovered in Azotobacter which in the presence of Mg^{2+} Protects the Fe-protein against the damage by oxygen by forming in its oxidized form a 1:1:1 teanary complex with the Fe-Protein and Mo-Fe Protein. The Protective Protein has a low molecular weight (14 kd - 24 kd) 2 Fe-2s Protein. This Protein is also known as 'Shethna Protein' or 'Fe-s Protein 2'.

⇒ Alternative Nitrogenases :

In some species of Azotobacter, Clostridium and Anabaena, alternative nitrogenase which lack molybdenum have been discovered.

Two types of Alternative nitrogenase are known one. Contains vanadium in the Place of Mo and is hence referred to as nitrogenase V or nitrogenase 2. The Second type of alternative nitrogenase neither contains Mo nor V and is referred to as nitrogenase 3. The alternative nitrogenases differ from nitrogenase 1 in physical, Chemical and biological Properties. The alternative nitrogenase can reduce acetylene to ethylene and further to ethane, Whereas nitrogenase 1 can reduce acetylene to ethylene only.

⇒ Mode of Action of Nitrogenase

Nitrogenase is a complex enzyme because it is a multi component System (Fe-Protein, MoFe-Protein ATP, Mg^{2+} are Required), which continues to work even in the absence of dinitrogen. The overall Reaction can be summarized as follows:

→ Acceptance of electrons by Fe-Protein from an electron donor

→ Subsequent reduction of MoFe-Protein which involves hydrolysis of Mg ATP

→ Reduced MoFe-Protein then reduces the substrate Mg-ATP binds to the Fe-Protein Prior to its reduction. The Fe-Protein which is a homodimeric dimer, binds 2 Mg-ATP and accepts electrons from a reduced ferredoxin (Fd) or flavodoxin (Fld). The Mg-ATP activated reduced Fe-protein then associates with the Mo-Fe Protein, forming a 'nitrogenase complex'. Two binding sites for Fe-Protein exist on the Mo-Fe Protein, but for the activity atleast one must be occupied. Thus the active complex can be 1:1 or 2:1 (Fe-Protein : Mo-Fe-Protein). When the complex is formed the electron is transferred from Fe-S cluster of the Fe-Protein to the Fe-atoms of FeMoco, and the Mg-ATP is hydrolysed. The nitrogenase complex then dissociates and the electron from the Fe atoms in the Mo-Fe Protein is utilized to reduce the substrate bound to the Mo-Fe Protein. As one electron is transferred from Fe-Protein to Mo-Fe protein at a time, a number of these electron transfer steps will be required before the final reduced product (H_2 and NH_3) are released from the enzyme.

Now, there is some evidence which suggests that the nitrogenase enzyme conducts a sequence of reductive protonations consistent with the

Known Chemistry of transition Metal-dinitrogen complexes. A scheme for reductive protonations has been proposed by Chatt, Dilworth and Richards

Fig: <u>A Scheme for Nitrogenase Activity</u>

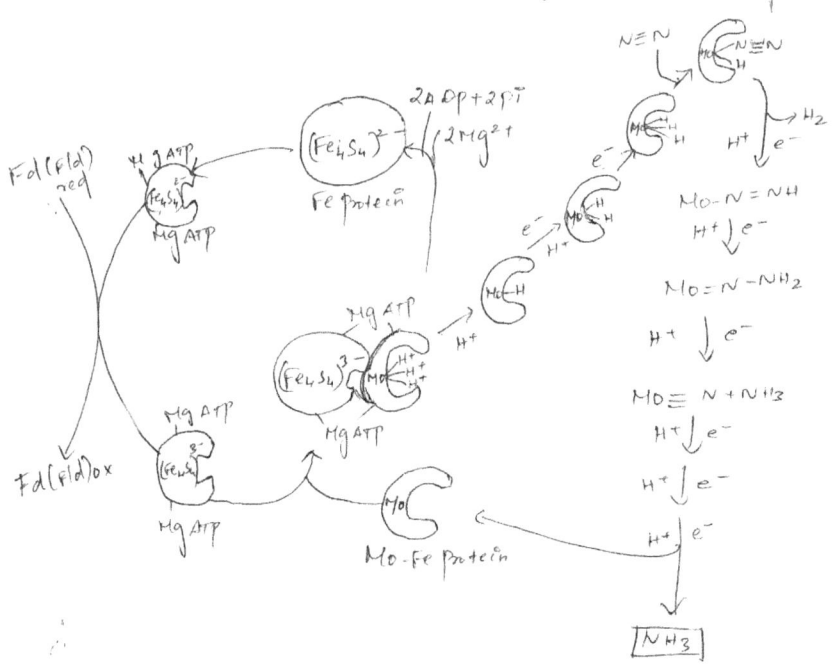

According to this scheme, dinitrogen is end bonded to a molybdenum atom of Mo-Fe Protein, electrons are transferred singly, and the 2 NH_3 molecules formed are released sequentially, rather than simultaneously.

Nitrogenase is a very sluggish and inefficient

enzyme, the main rate limiting step being the dissociation of Fe and MoFe Proteins following each electron transfer, and the enzyme Must go through several catalytic reduction cycles before the appearance of the Product (NH₃). The turnover time is 185 ms/electron pair at 23°c. The rate of turnover is also dependent on many other factors including substrate concentration and exchange reactions between Mg ATP and Mg ADP on the Fe-Protein.

⟹ <u>Role of Mg ATP</u>

A minimum of 2 Mg ATP molecules are hydrolysed for each electron transferred from reduced Fe Protein to MoFe-Protein. The hydrolysis of Mg ATP helps in pushing equilibrium of the reaction to the right. The binding of Mg ATP induces a conformational change in the Protein. Due to this conformational change, the environment of Fe-S cluster is altered, increasing the energy of the electron as well as sensitivity to Oxygen. Though the Primary function of Mg ATP is in the electron transfer between 2

Components of nitrogenase, some experiments suggest that it may have an additional function concerning substrate binding.

The ATP needed for nitrogen fixation can be supplied in different N_2 fixers as follows:

→ In the aerobic organisms and nodule bacteroids: by the oxidation of respiratory substrates

→ In the anaerobic Clostridium: via the 'phosphoroclastic' cleavage of Pyruvate to acetate

$$Pyruvate + CoASH + Fd_{ox} \xrightarrow[\text{dehydrogenase}]{\text{Pyruvate}} Acetyl\ CoA + CO_2 + Fd_{red}$$

ATP Generating system

$$Acetyl\ CoA + Pi \xrightarrow{\text{Transacetylase}} Acetyl\ phosphate + CoASH$$

$$Acetyl\ phosphate + ADP \xrightarrow{\text{Acetokinase}} Acetate + ATP$$

→ In Photosynthetic Systems : Via photophospho -rylation

⇒ Generation of reducing power and Pathway of electron Transfer :

In addition to ATP, reducing Power is also required in the Process of biological nitrogen fixation. The electron donors must have a very low, negative redox potential. The ferredoxins and flavodoxins are strong reducing agents, and are involved in biological N_2 reduction. They are present in all diazotrophs. In <u>clostridium pasteurianum</u>, the ferredoxin is generated via the 'phosphoroclastic' cleavage of the Pyruvate. In addition ferredoxin also transfers electrons directly from hydrogenase to CP2:

$$H_2 \longrightarrow Hydrogenase \longrightarrow Ferredoxin \longrightarrow CP2.$$

In Azotobacter <u>chroococcum</u> ferredoxin is present but a flavodoxin (azotoflavin) is the Primary reductant for nitrogenase. This is because semiquinone (half-reduced) form of flavodoxin is usually stable to oxidation by air and would be more suited to aerobic way of life of the bacterium.

In cyanobacteria, photosynthetically reduced NADP or dark-produced NADPH (via Pentose Phosphate Pathway) donates electron to nitrogenase via ferredoxin. In heterocysts of cyanobacteria that lack photosystem II, photosynthates are imported from the vegetative cells and metabolized according to the following sequence in order to provide reductant.

$$Pentose\ phosphate \xrightarrow{e^-} NADP \xrightarrow{e^-} Fd \xrightarrow{e^-} Nitrogena$$
$$cycle\ intermediates$$

with Pyruvate $\xrightarrow{e^-}$ (above Fd)

ultimate electron donors are photosynthates in phototrophs, Carbon substrates in heterotrophs and presumably election donors such as H_2 or Fe^{2+} in chemotrophs.

⟹ Substrate - reducing sites

Nitrogenase has broad substrate specificity and catalyses the reduction of a variety of substrates. All reductions by nitrogenase require 2 or multiple of 2 electrons. Dinitrogen and Proton (presumably from water) are the natural substrates. The enzyme involves H_2 whether or not a reducible substrate is added

Table → Substrates reduced by Nitrogenase

Name	Major products	Electrons
Dinitrogen	$2NH_3$	6
Proton	H_2	2
Acetylene	$H_2C=CH_2$	2
Azide	$N_2 + NH_3 + N_2H_4$	2
Nitrous oxide	$N_2 + H_2O$	2
cyanide	$CH_4 + NH_3$	6
Alkyl cyanides	$RCH_3 + NH_3$	6
Alkyl isocyanides	$RNH_2 + CH_4$	6

Nitrous Oxide, azide, acetylene, cyanide and methyl isocyanide not only inhibit N_2 reduction but also serve as substrates for nitrogenase. Carbon monoxide (CO) is a non-competitive inhibitor of N_2 fixation and H_2 is a competitive inhibitor. CO inhibits reduction of all substrates except Protons. $Mg ADP$, which is a product of nitrogenase reaction, inhibits the activity of the enzyme.

⟹ Nature of Active site

Considerable evidences suggest that the active substrate reducing sites of nitrogenase contain transition Metal atoms (MO, Fe) which are oxidised and reduced during catalysis, and to which reducible substrate binds.

⟹ Bound Intermediates

The first stable product of dinitrogen reduction is ammonia. It is now generally accepted that bound intermediates are produced during the reduction of N_2 to ammonia. Based on studies with dinitrogen complexes of transition Metals Mo and W, the likely structure for the intermediate is

$$MO \overline{\cdots} N \overline{\cdots} NH_2$$

(or)

$$MO \overline{\cdots} N \overline{\cdots} NH_3^+.$$

B. Ammonia Assimilation and Transport

The first stable product of biological nitrogen fixation catalyzed by the enzyme nitrogenase complex is ammonia (NH_4^+). In free-living diazotrophs, ammonia is utilized for the synthesis of amino acids. In symbiotic nitrogen fixation, ammonia is excreted into the host cytoplasm, where it is assimilated into organic molecules in the form of amino acids, amides or ureides, which are transported through xylem sap and made available for plant growth. In leguminous plants, the main nitrogen transport compounds are amides (asparagine and glutamine) and ureides (allantoin and allantoic acid). The other compounds citrulline and 4-methylene glutamine are observed in a few species.

→ Ammonium Assimilation

The first organic compound formed after ammonia assimilation is glutamine. The reaction is catalysed by the activities of glutamine synthetase (GS) and glutamate synthase (GOGAT : glutamine oxoglutarate aminotransferase). These enzymes remain repressed in bacteroids and induced in host cells. GOGAT exists both in cytoplasm and plastids. The reaction are as follows:

$$\text{Glutamate} + NH_4^+ + ATP \xrightarrow[Mg^{2+}]{GS} \text{Glutamine} + ADP + Pi$$

$$\text{Glutamine} + 2\text{-Oxoglutarate} + NAD(P)H + H^+ \xrightarrow{GOGAT}$$
$$2 \text{ Glutamate} + NAD(P)^+$$

The net reaction is:

$$2\text{-Oxoglutarate} + NH_4^+ + ATP + NAD(P)H + H^+ \longrightarrow$$
$$\text{Glutamate} + NAD(P)^+ + ADP + Pi.$$

Another reaction for NH_4^+ assimilation is catalysed by the enzyme glutamate dehydrogenase (GDH):

$$2 \text{ oxoglutarate} + NH_4^+ + NAD(P)H + H^+ \xrightleftharpoons{GDH}$$
$$\text{Glutamate} + NAD(P)^+ + H_2O$$

The nitrogen incorporated into the amide group of glutamine is further used for the synthesis of α-amino acids and nitrogen transport compounds.

The Pathway of ammonium assimilation is altered in the presence of inorganic nitrogen. The level of organic nitrogen in the xylem sap decreases with the increase in inorganic nitrogen.

⟹ Synthesis of Amides

Asparagine and glutamine are the major nitrogen-transport compounds present in xylem sap of several legumes. Asparagine is probably superior to glutamine for N transport. The synthesis of asparagine from aspartate takes place by glutamine dependent asparagine synthetase (AS).

Aspartate + Glutamine + ATP $\xrightarrow{AS}$ Asparagine + Glutamate + ADP + pi.

The aspartate required for continuous synthesis and export of asparagine comes from the transamination of oxaloacetate by glutamate dependent aspartate aminotransferase.

⮞ Synthesis of Ureides

Ureides are the nitrogen transport compounds in many tropical and subtropical legumes. The major ureides are allantoin and allantoic acid. In few plants, citrulline is also used for nitrogen transport. Allantoin and allantoic acid are energetically the most efficient nitrogen transport compounds.

Citrulline synthesis takes place from ornithine and carbamoyl phosphate. The enzyme Carbamoyl phosphate synthetase (CS) catalyses the synthesis of Carbamoyl phosphate from NH_4^+ of glutamine and CO_2

Glutamine + CO_2 + 2ATP $\xrightarrow{CS}$ Carbamoyl Phosphate + Glutamate + 2ADP + pi

The synthesis of ornithine takes place from glutamate and acetyl co-enzyme A.

Figure: Ornithine cycle

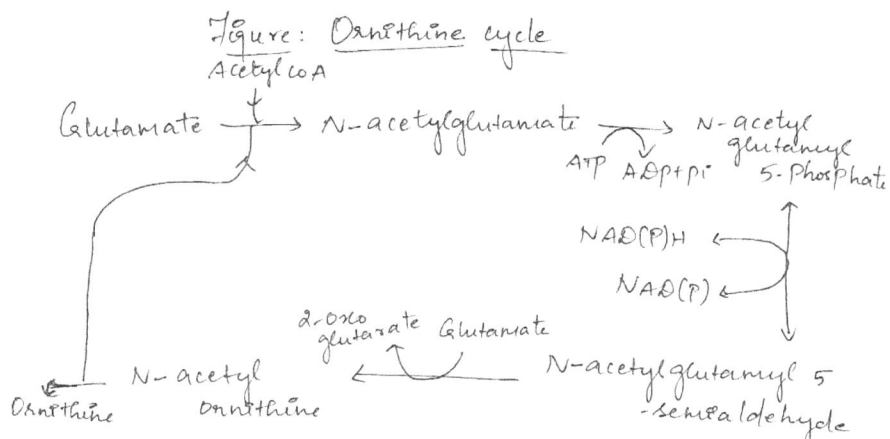

The synthesis of citrulline from ornithine and Carbamoyl phosphate is catalysed by ornithine carbamoyl transferase (OCT).

Ornithine + Carbamoyl $\xrightarrow{\text{OCT}}$ Citrulline + Pi
Phosphate

Allantoin and allantoic acid are formed by the oxidative catabolism of purines. This is supported by the presence of several fold higher levels of enzymes involved in purine catabolism such as xanthine dehydrogenase, uricase, allantoinase in nodules of ureide exporting legumes as compared with amide exporting legumes. Allopurinol, an irreversible inhibitor of xanthine dehydrogenase also decreases the level of ureides in nodules and xylem sap. This shows that the ureides are formed by the oxidative catabolism of Purines. Several enzymes involved in Purine and ureide biosynthesis have been purified and characterized from nodules. Inosine monophosphate is the intermedi Metabolite in de novo pathway of purine synthesis, which serves as the precursor of ureides. The precursors of various atoms of a purine nucleus a the pathway for the synthesis of inosine monophospha is shown in the figure:

$$\boxed{\text{Ribose - 5 - phosphate}}$$

$$\text{ATP} \searrow \text{Phosphoribosyl pyrophosphate synthetase}$$
$$\text{AMP} \nearrow$$

$$\boxed{\text{5 - Phosphoribosyl - 1 - pyrophosphate}}$$

$$\text{Glutamine} \searrow \text{Amidophosphoribosyl transferase}$$
$$\text{Glutamate} \nearrow$$

$$\boxed{\text{5 - Phosphoribosyl - 1 - amine}}$$

$$\text{Glycine + ATP} \searrow \text{Glycinamide synthetase}$$
$$\text{ADP + pi} \nearrow$$

$$\boxed{\text{Glycinamide Ribonucleotide}}$$

$$\text{Methenyl tetrahydrofolate} \searrow \text{Glycinamide formyl transferase}$$
$$\text{Tetrahydrofolate} \nearrow$$

$$\boxed{\text{Formylglycinamide Ribonucleotide}}$$

$$\text{ATP + Glutamine} \searrow \text{Formylglycinamide synthetase}$$
$$\text{ADP + Pi + Glutamate} \nearrow$$

$$\boxed{\text{Formylglycinamidine Ribonucleotide}}$$

$$\text{ATP} \searrow \text{Aminoimidazole ribonucleotide}$$
$$\text{ADP + pi} \nearrow \text{synthetase}$$

$$\boxed{\text{5 aminoimidazole ribonucleotide}}$$

$$\text{ATP + CO}_2 \searrow \text{Aminoimidazole ribonucleotide carboxylase.}$$
$$\text{ADP + pi} \nearrow$$

$$\boxed{\text{5 aminoimidazole 4 - Carboxy ribonucleotide}}$$

$$\text{Aspartate + ATP} \searrow \text{Synthetase and Succinate lyase.}$$
$$\text{Fumarate + ADP + Pi} \nearrow$$

$$\boxed{\text{5 - aminoimidazole - 4 - Carboxamide Ribonucleotide}}$$

$$\text{formyl tetrahydrofolate} \searrow$$
$$\text{Tetrahydrofolate} \nearrow$$

$$\boxed{\text{5 - formamidoimidazole - 4 - Carboxamide ribonucleotid}}$$

$$\text{H}_2\text{O} \nearrow$$

$$\boxed{\text{Inosine - 5 - monophosphate}}$$

Catabolism of Inosine monophosphate leads to the formation of allantoin and allantoic acid in a sequence of steps:

Inosine monophosphate

NAD
NADH → Inosine monophosphate dehydrogenase

↓

Xanthosine Monophosphate

Pi → Nucleotidase

↓

Xanthosine

Ribose → Nucleosidase

↓

Xanthine

NAD
NADH → Xanthine dehydrogenase

↓

Uric acid

$CO_2 + H_2O_2$ → Uricase

↓

Allantoin

↓ Allantoinase

Allantoic acid

Transport of Nitrogen compounds

Tropical legumes are ureide transporters and temperate legumes are amide transporters. Nodules of ureide transporters are of determinate type, with closed vascular systems, which facilitates

higher rate of water flow through nodules.
The ureides are nearly 9 times less soluble than
amides. Thus, the high rate of water flow through
determinate nodules keeps ureide concentration below
crystallization during transport. In amide transporting
plants, the nodules are indeterminate with open
and branched vascular system which restricts water
flow. Under restricted water flow of indeterminate
nodules, high solubility of amides facilitates their
transport under varying transpiration conditions. In
Vicia, Trifolium and other amide exporting nodules
the level of asparagine and glutamine increases in
nodules during night or when the humidity in
the atmosphere is high. Due to high solubility
of amides their crystallization is prevented
under such conditions

Chapter-6

Physiological Aspects of Nitrogen Fixation

A. Protection of Nitrogenase against O_2:

MoFe Protein and Particularly Fe-protein is damaged by oxygen. Therefore, all diazotrophs are obliged to protect their nitrogenase enzyme system from damage of O_2. The Mechanisms that have been evolved by aerobes to Protect nitrogenase are

→ Respiratory Protection in Azotobacter
→ Conformational Protection in Azotobacter
→ Gum Production
→ heterocysts in cyanobacteria
→ Possession of superoxide dismutase and catalase

Respiratory Protection in Azotobacter

Since Azotobacter is a N_2 fixing obligate aerobe, there should exist some Mechanism to prevent the access of O_2 to the oxygen-sensitive sites of nitrogenases. The protective function in Azotobacter is performed by exceptionally high respiration rate, backed up by conformational Protection Azotobacter has a specialized respiratory system adapted to respiratory Protection. Populations growing at high PO_2 will develop a matching respiratory activity as well as a new terminal oxidase (cytochrome d).

Azotobacter Vinelandii has a branched respiratory chain, one of which is operative at low pO_2 with all the 3 phosphorylation sites functioning to yield a high P:O ratio. In other pathway, which is operative at high pO_2 Phosphorylating sites I and III are lost and the activity of site II is reduced resulting in a low P:O ratio. At high pO_2 cytochrome b→d pathway constitutes the major terminal route to O_2 and does not conserve energy. Thus high respiration rate scavenges O_2, preventing the penetration of O_2 to the N_2-fixing site.

Figure: Branched Respiratory chain in Azotobacter Vinelandii

I, II, III are the approximate sites of phosphorylation of ADP to ATP.

☞ Conformational Protection in Azotobacter

   Nitrogen fixation is inhibited when Azotobacter is suddenly exposed to an O_2 concentration higher than the concentration at which it has been grown. When oxygen concentration is lowered, nitrogen fixation begins quickly again. A phenomenon in which nitrogenase activity becomes reversibly switched on or off in response to decreased or increased pO_2 is known as conformational protection. This protection mechanism operates when the O_2 level becomes too high for respiratory protection to cope with. 'Switch on' is the reuse of pre-existing nitrogenase and not the synthesis of new enzyme whereas during 'switch-off', when pO_2 is high, nitrogenase proteins are not damaged but protected.

The plausible view of the phenomenon is that, the oxidized form of 2Fe2S protective protein, in the presence of Mg^{2+}, binds to the Fe-protein and the Mo-Fe protein, forming a 1:1:1 ternary inactive complex. On complex formation the oxygen sensitive sites are protected from the damage of oxygen.

When O_2 level is lowered, it allows the formation of reduced species of protective protein and its consequent dissociation from the nitrogenase proteins. As a result of this the enzyme becomes active again. Thus a conformational protection is caused by passive protein-protein interaction to screen oxygen sensitive sites.

It has also been suggested that the oxidised 2Fe2S protective protein might regulate nitrogenase activity by associating and reversibly inactivating the Fe-Protein. Reactivation occurs by reducing or removing 2Fe-2S protein. In this way the Protective protein can both regulate nitrogenase activity and Prot against damage by oxygen.

➤ Gum Production:

Most of the diazotrophic bacteria, specially _Derxia gumniosa_ and _Xanthobacter flavus_. Produce gum (polysaccharide). Gummy material offer some passive protection against O_2 by decreasing oxygen solubility thereby impeding O_2 uptake into the cell.

➤ Heterocysts in cyanobacteria

In free living filamentous cyanobacteria, heterocysts which are enlarged cells, occur at regular intervals, along the filaments, are the principal sites of nitrogenase activity. Heterocysts have thicker cell walls than vegetative cells. They prevent the entry of O_2 into the cell by offering a resistance to the diffusion of O_2. Heterocysts also lack Photosystem II, which is responsible for oxygen evolution. Thus heterocysts act as biological compartment.

In unicellular cyanobacteria such as _Gloeothece_, nitrogenase is protected by temporal separation of diazotrophy and photosynthesis. Most of the nitrogen

is fixed in the dark at the expense of Carbohydrate
reserves built up by Photosynthesis during the day.

→ Catalase and Superoxide Dismutase

Oxygen metabolism produces superoxide radicals
which may interact with nitrogenase proteins and
damage its activity. These can be detoxified by a
combination of superoxide dismutase which converts
O_2 to H_2O_2 and catalase

B. Role of Leghaemoglobin

Leghaemoglobin is a monomeric protein with high
affinity for oxygen. It binds O_2 reversibly and transports
it to the ATP generating machinery of the bacteroids
necessary for the reduction of dinitrogen.

Leghaemoglobin is a haemoprotein synthesised in the
nitrogen fixing nodules by the co-operative action of both
Partners - globin protein is synthesised by the host plant
and haem cofactor by the bacteroids. It is
restricted to the infected cells within nodules where
it constitutes 25-30% of the total soluble protein
of the cell.

The Physiological roles played by Leghaemoglobin
in nodules include.

→ facilitate O_2 diffusion across O_2 free layer
surrounding the bacteroid

→ acts as an O_2 buffer to allow the correct concentration of O_2 to the bacteroid surface to support ATP synthesis and nitrogenase activity. Within the nodul leghaemoglobin is about 20% oxygenated. The nodule bacteroids have 2 terminal oxidases, each having high and low affinity for oxygen, to accept O_2 from oxyleghaemoglobin. Terminal oxidases with high affinity for O_2 operates at low O_2 concentration and is tightly coupled to ATP production. Leghaemoglobin facilitates O_2 flow at low O_2 concentration to the high-affinity oxidase system. The low affinity oxidase may serve to protect nitrogenase from excess O_2.

c. Regulation of Nitrogenase Activity

As per the need, the synthesis and activity of nitrogenase and other proteins required in diazotrophy are regulated.

➤ Regulation by NH_4^+

A 'switch-off' 'switch-on' Mechanism controlled by NH_4^+ concentration seems to be acting in some photosyn-thetic bacteria (Rhodospirillum and Rhodopseudomonas). The level of NH_4^+ required for the switch-off of activity is generally much lower than that required for repression of enzyme synthesis. Ammonium inactivates the Fe-protein by covalent modification. There is no direct effect of NH_4^+ on purified nitrogenase proteins. But glutamine, an assimilatory product of NH_4^+ has been implicated as a trigger molecule in this system. The modifying group of Fe-Protein, adenosine diphosphoribose binds to arginine of the Fe-Protein. The modified Fe-Protein can still undergo oxidation or reduction and also binds MgATP, but is incapable of electron transfer to MoFe-Protein. In these photosynthetic bacteria for switch-on of nitrogenase activity, Mn^{2+} dependent enzymic activation of ADP-ribosyl Fe-Protein is required.

→ Regulation by oxygen

Since nitrogenase functions under strictly anaerobic conditions, many systems have been devised to regulate the supply of oxygen. In addition to NH_4^+ and O_2, nitrogenase activity is also regulated by Mo and energy supply. The primary role of Mo is in the regulation of nitrogenase synthesis. Similarly in blue green algae, under conditions of Mo-deficient de-repression, nitrogenase synthesis takes place but MoFe-Protein is inactive. In all such cases, addition of Mo to the culture restores the nitrogenase activity. Nitrogenase activity is also regulated by MgATP:ADP ratio.

Chapter - 7

Nodulation - Early and Late Events

The establishment of legume-Rhizobium symbiosis is a complex process involving physiological and biochemical properties of both the bacterium and host plant. There is interaction of a particular legume species with its respective Rhizobium symbiont which is known to be fairly specific which will lead to the establishment of cross-inoculation groups.

The events involved in the process of root-hair infection are recognition of plant host and Rhizobium species, rhizobial adherance to root hairs, root-hair curling, root-hair infection, root nodulation and transformation of vegetative bacteria into enlarged pleomorphic bacteroids which fix nitrogen.

Recognition

The early steps of infection involve cell-cell contact and recognition by which plant and bacterial signals are exchanged. Various models prepared related to this implicate plant lectins (glycoproteins) and surface carbohydrates (exopolysaccharides) of rhizobia in the recognition process. The host plant lectins located on the root surface recognize carbohydrate receptors on the compatible Rhizobium cell surface and hence bind the bacteria to the root.

According to another hypothesis, host induced alterations of capsular polysaccharides of Rhizobium and de-novo synthesis of specific proteins by both the partners in response to each other, appears to be the first step in the series of signal and response interaction.

Root hair wall, Clover lectin — R. trifolii — Saccharide Receptor

⇒ Infection

The next stage after recognition, is the entry of rhizobia into the host through the root hair, which gets deformed due to hormone like substance produced by the bacteria. The root hairs become elongated, curled and branched. An interchange of small diffusal molecules between the host and the bacterium may occur before the deformation takes place. The curling of the root hair may facilitate the infection as the bacteria become enclosed by the root-hair walls and the hydrolysing enzymes may not be allowed to diffuse out. It seems that the cell-wall is altered at a localised site by hydrolytic enzymes which can either be from host (pectinases, cellulases, glucan hydrolases, various glucosidases) or from the rhizobia (pectinases, hemicellulases, cellulases) or both.

Plant flavones and flavonones are found to act as inducers of Rhizobium nodulation (nod) genes which specify early events like root hair curling and nodule induction.

⟹ _Formation of infection Threads_

Infection threads are tubular structures that carry Rhizobium cells, from Root surface into the Root cortex. The formation of infection threads in root hair cells provides the means of entry for rhizobia, in most of the legumes. In groundnut infection thread is not formed, and the bacteria enter the root at the junction of the root hair and the epidermal cells. On the entry of rhizobia into root-hair cell, a new cell wall is formed cutting the bacteria off from the contents of the host cell. The apparent continuity of the infection thread wall with the original root hair cell is due to the synthesis of a new wall layer which is fibrillar containing pectin and cellulose and no callose. The bacteria divide in the enclosed space of the infection thread which extends to grow towards the base of the root hair cell.

Rhizobia enter other cortical cells intercellularly through cell wall middle lamella. The cell wall is structurally altered and degenerates. The degenerated material surrounds the rhizobia in the middle lamella and intercellular zooglea are formed.

→ Nodule Development:

The infection thread continues to grow beyond the root hair cell and penetrates the cortex of root. The thread crosses the already existing cell walls. The rhizobia produces gets embedded in a mucopolysaccharide containing matrix, termed Zoogleal matrix. Extensive branching of the thread occurs within the cortex, resulting in the infection of many cells by the same thread. The rhizobia are released shortly after the penetration of the cortical cells by the infection thread. The disintegration of thread wall takes place possibly due to Rhizobium induced pectinase activity, and cellulase from plant cell. The rhizobia embedded in zooglea enter into new empty cells through dissolution of cell wall.

Vesicles derived from the infection thread Membrane are formed at the sites where disintegration has occurred. From these vesicles, rhizobia embedded in zoogleal matrix and surrounded by a Membrane, termed bacteroid Membrane, are budded off by a process identical to endocytosis. Initially the peribacteroi.

Membrane is formed from the infection thread Membrane.
Nodule-inhabiting swollen and/or distorted forms of
rod-like rhizobia are termed as bacteroids.

→ Biochemical changes During Nodule maturation

With the invasion by rhizobia, the plant cells
undergo rapid cell division. Diploid cells divide and
become tetraploid. As the infected cells mature, the
bacteroids cease dividing and increase 10-40 fold in
volume. Bacteroid shapes undergo progressive Modification
In addition to morphological and cytological changes,
a series of biochemical changes occur in the plant
cells and the rhizobia present in the nodule. To cope
up with the low oxygen concentration within the
nodule, bacteroids have an altered set of cytochromes
c-552, P-420 and P-450.

The nodule environment induces the synthesis of
bacteroid-specific proteins and also nodule specific
proteins, nodulins. There is a Massive synthesis of
Leghaemoglobin

Chapter -7

Molecular Biology of Nitrogen Fixation

Symbiosis between rhezobia and legume results from a cell to cell interaction which leads to the formation of nodules. Various bacterial genes involved in symbiosi are nif (nitrogen fixation), fix (ability to fix nitrogen) and nod (ability to nodulate).

→ **Plant Genes (Nodulin genes)**

The plant genes which take part in symbiotic nitroge fixation can be grouped into 3 categories:

a) Genes whose expression is induced

b) Genes repressed

c) Constitutively expressed genes.

The products of inducible genes or nodule specific host proteins are known as nodulins. These are not found in normal root tissue. Nodulins are divided into 2 groups:

1) C-nodulins common to all legumes

2) S- nodulins present in specific species

Based upon their functions, nodulins are grouped into 3 categories:

a) nodulins responsible for nodule structure

b) nodulins responsible for enzymatic activities of nodule

c) nodulins involved in nitrogen fixation.

Nodulins are numbered either according to their molecular weight (eg:- nodulin-35 has 35 kd molecular weight) or by their order of size (Number one (N-1) being the largest polypeptide and N-30 (globin) the smallest one]. Nodulins can be identified either by nodule specific poly A-m RNA or by immunological assay using antisera prepared against the total nodule proteins. Among various nodulins, leghaemoglobin is the most well characterized nodulin. At present about 30 nodulins are known and functions of most of these is not yet understood.

Rhizobial Genes

The ability of rhizobia to invade legume plant and stimulate the host to develop nodules depends upon several genes. Several bacterial genes are involved in nodule development. Bacterial genes concerning symbiosis are collectively referred to as `Sym'genes which consists of 'nod', 'nif' and 'fix' genes.

NOD Genes :

Bacterial genes involved in nodulation are collectively referred to as nod genes. Several bacterial mutants defective in nodulation have been characterized to understand the various steps of nodule formation. They are,

   roc - root colonization

   roa - root adhesion

   hab - hair branching

   had - hair deformation

   hac - hair curling

   hsn - host specificity of nodulation

   inf - infection

   noi - nodule initiation

   inb - infection thread branching

   bar - bacterial release

   bad - bacteroid development.

     Based on their phenotypic properties, nod genes can be grouped into 2 broad categories:

     a) nod genes whose defect can be complemented by heterologous <u>Rhizobium</u> strains

     b) nod genes whose defects cannot be complemented by genes from a heterologous <u>Rhizobium</u> species

   nod A, B, C → Genes belonging to first category and are necessary for root hair curling

   nod D → regulatory gene that is needed for the activation of transcription of other nod genes.

The nod genes belonging to the second category are also known as hsn (host specificity of nodulation) Two of the hsn genes, hsn A and hsn B (called nod F and E) are homologous among R. <u>trifolii</u>,

R. meliloti and R. leguminosarum.

hsn C and D (nod G and H) — also implicated in host range specificity in R. Meliloti

In addition to these, there are 5 nod genes I, J, L, M and X, identified using DNA sequence analysis but their exact function is unknown.

Fix Genes:

Genes controlling the ability of rhizobia to fix nitrogen within a nodule are referred to as fix genes.

Nif Genes.

The genes needed for nitrogen fixation are designated as :nif genes.

In Klebsiella pneumoniae, the nif genes lie between the his gene that codes for histidine synthesis and She A gene coding for the synthesis of Shikimic acid. The nif cluster is composed of 21 genes located on 23 Kb DNA. The region is composed of 9 transcriptional units of which 7 operons are transcribed towards his locus and 2 towards Shi A locus.

Nif genes code for nitrogenase polypeptides, components of electron transport to nitrogenase and factors involved in genetic regulation of nif transcription

→ α and β subunits of nitrogenase component 1 are coded by nif D and K

→ component 2 polypeptide is coded by nif H gene

→ Mutations in nif E, N, Q, B, S, V and U genes affect the activity of MoFe-Protein

→ nif V mutants can reduce acetylene

→ nif B, N, E are needed for the synthesis of FeMo-co

→ nif Q involved in Mo uptake

→ nif S and V are believed to be involved in the modification of MoFe-Protein.

→ For Production of component 2 genes nif H, M and Z are required

→ nif M mutants results in an inactive Fe-protein.

→ nif A and L genes have a regulatory effect on other nif genes.

    Expression of nif genes is regulated by

1) ntr genes which also regulate a large set of genes that are involved in the transport, degradation and assimilation of nitrogen compounds

2) 2 nif products coded by nif A and L genes.

REFERENCES

→ Annie Ragland, Rajakumar, Rajarathnam, Jayakumar. (2011). Plant Physiology. Saras Publication

→ V. K. Jain (2011). Fundamentals of Plant Physiology. S. Chand Publishing.

→ Evans, H. J and Nason A. 1953. Pyridine nucleotide - nitrate Reductase from extracts of higher plants. Plant Physiol. 28: 233 - 254

→ Nason A. 1956. Enzymatic steps in the assimilation of nitrate and nitrite in fungi and green plants. 109-136. In. W. D. McElory and H. B. Glass (eds)., Inorganic nitrogen Metabolism. John Hopkins Press., Baltimore, Md.

→ Nason A and Evans H. J (1954). The phosphopyridine nucleotide - nitrate Reductase in Neurospora. J. Biol. Chem. 202: 655

→ Jody Barley. Recent Advances in Plant Biochemistry. Random Publications. 92B2220704 (ISBN)

* * * * *